Ghee

A Guide To The Royal Oil

Kathryn S. Feldenkreis

Library of Congress Card Catalog Number: 96-90831
ISBN 978-1-57502-354-0

Printed on Recycled Paper

Printed in the USA by
Morris Publishing
Kearney, NE

DEDICATION

*This Book is Dedicated to the
Healing Power
That resides in all foods
made with Love*

ACKNOWLEDGMENTS

I wish to extend a heartfelt thank you to my husband, Howard, who is the best partner in business and life. To my parents, Ed and Helen Otten. Also, to my dear friends Eleanor Hunter, Roger and Susan Thicke and Dr. Bob Yudin—for the encouragement and inspiration that you have always given me. To Connie Newton for teaching me the value of service. For all of your editing and typesetting, thanks to Jeanette Bogren and Diane Baker. Gratitude also goes to the people at Organic Valley Dairy Co-operative for their dedication to the highest standards in organic farming and sustainable agriculture. To Daisy Chisholm at Forbes Graphics, thank you for your creative input and help on the cover photograph. Thank you, especially, to all the people who have purchased Purity Farms Ghee over the years. You have made it all possible.

Go, mark the matchless working of the Power
That shuts within the seed of the future flower.

—*Cowper*

CONTENTS

Teach us delight in simple things,
And mirth that has no bitter springs.

—*Kipling*

PREFACE

This book is intended to be a household guide to preparing and cooking with ghee. In order to sell ghee, or any other food product, you must approach your state health inspector and go through the required licensing and inspection procedures.

Any therapeutic uses for ghee mentioned in this book are for informational use only and are not intended to replace the advice of a health care professional.

ENJOY THIS BOOK IN GOOD HEALTH!

The art of dining well is no slight art,
the pleasure not a slight pleasure.
—Michel Eyquem De Montaigne
16th Century French Essayist

INTRODUCTION

In 1983, my husband, Howard, and I were living in Evanston, Illinois. One evening, a friend brought us Xeroxed copies of pages from an Ayurveda manual. We had heard a little about Ayurveda and knew that it was the ancient system of natropathic medicine in India. With a cup of tea in hand, we began to read these obscure manuscripts with a curiosity that quickly grew into fascination. The Ayurvedic approach to health was about establishing balance, with careful attention to diet, herbal preparations and behavioral modification. That very evening, we set out to incorporate some of these new-found principles into our lives. And that is when I first discovered the importance of ghee (clarified butter). Ayurveda holds ghee in the highest regard. It is mentioned as a remedy for a myriad of conditions from insomnia to burns and even senility.

Since I have had an aversion to butter since childhood, I assumed the claims about ghee were at best over-rated or possibly just folk tales. In my enthusiasm to jump into the study of Ayurveda, I purchased a few pounds of butter and made my first batch of ghee. Clarifying the butter was a surprisingly gratifying experience. Intuitively, I seemed to know when the solids were about to separate and if the heat was too high. But the most satisfying was the moment when the butter magically transformed into ghee—Eureka! Now, if I could only get used to eating it. Boy, was I surprised. The tiniest drops of ghee gave a special aroma and flavor to every food.

I began to make more and more of this golden liquid, to the point that people in our apartment building would tell me that on the days I made the ghee the wing of our building would smell like popcorn or freshly baked cookies. I was garnering a small group of people who asked me to make ghee for them on a regular basis.

By now we were living in Silver Spring, Maryland, and I would imagine myself making ghee in huge quantities and selling it all over the country. Far fetched as it was at the time, I continued to refine my ghee-making techniques and research outlets for this esoteric product.

My aspirations took a quantum leap when the secretary at a nearby school encouraged us to rent their school kitchen which was fully equipped, yet sitting vacant. It was perfect for our needs and we began to manufacture ghee on a large scale. That was the genesis of Purity Farms, Inc. Ghee Company. Fourteen years and millions of jars later, Howard and I consider ourselves experts on ghee. This simple product has been a catalyst for many extraordinary lessons—most notably all the experiences that come with developing and expanding a business. The actual process of making ghee and its many uses holds the same fascination for us today that it did when we first read about it.

I look forward to sharing with you all that I have learned as if you were visiting my home for a day of making ghee. Since this is a new oil in most households, I want to give you all the information you need to go about making and using ghee confidently.

Making ghee can be regarded as a ritual—a special time when we connect with that sense of careful, purposeful, almost devotional activity. It is a reminder of the sacredness of something so simple as clarifying butter. In a world of fast food, microwaves and instant meals, I invite you to enjoy the gracious art of making ghee.

The cows yield butter and milk inexhaustible
for thee set in the highest summit.
—*Rig Veda, IX, 2.7*

BEGINNING WITH BUTTER
Butter and Butter Processing

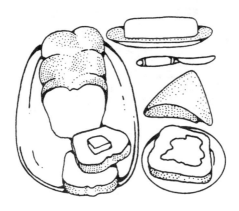

Ordinary butter that we purchase from the grocery is a composition of 80% butterfat, 16% water, 3% salt, and 1% milk fat solids or curd. The curd is made of proteins, mineral matter and lactose or milk sugar.

Conventional butter is churned from cream instead of milk, because cream has ten times more available butterfat. One would have to start with ten quarts of milk to yield enough cream to make one pound of butter.

The dairy states of Wisconsin and Minnesota have earned their reputations by producing over 425 million pounds of butter annually. Many people are surprised to hear that California's butter production is second only to Wisconsin. See the chart on page 6 for a breakdown of North America's leading butter producers.

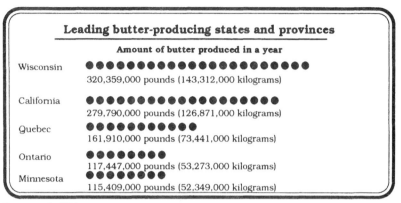

Figures are for 1990.
Sources: U.S. Department of Agriculture, Statistics Canada.

Worldwide, the Russians lead in butter production and con-sumption by a wide margin. Each year Russians will consume ten pounds of butter per capita. Compare this to approximately 4.5 pounds of butter per capita for every American.

Following the Russians in international butter production is France, then the United States, Germany and India. Butter is just a part of the dairy picture since it accounts, in the United States anyway, for utilizing only 18% of the milk produced. Over the counter fluid milk sales and cheese make up the biggest portions of dairy product sales.

For an industry of this magnitude it is surprising that the first large scale commercial creamery didn't pop up in the United States until 1856 in Orange County, New York. After that, creameries were established quickly throughout the country. By 1900 commercially churned butter was available in most stores coast to coast. Even with the availability of butter as a commodity, the processes were *at best slow and laborious—the butter produced in small batches by wooden churns. The Europeans leapt forward in butter production in the early 1880's due to an invention by Carl Gustav de Lavel, a Swedish engineer. His centrifugal separator automated the process of separating the cream from the milk. This quantum leap in efficient, modern butter production came as a result of the continuous process machine invented in Australia in 1937. This was dubbed the new-way process, which followed the Fritz method for churning developed in Germany in 1933. Variations on these central concepts were developed and improved upon all over the world.*

The key to the quality of any food product is in the raw material that one starts with, in this case the cow's milk. In all my years of making and selling ghee, I would invariably get calls from people who said the ghee tasted different in the summer than the winter. This, of course, is the result of the diets and grazing habits of the cows in the different seasons. The more fresh grass they eat, the more chlorophyll in their system and the more yellow the ghee. The taste of summer ghee is more potent in flavor, while the winter ghee is milder and a softer shade of gold. In conventional butter making, colors are often added to the butter to give it a consistency that is more desirable in the marketplace. More on this later, when I will discuss all the unnecessary additives put into this originally simple, pure product.

Butter is graded, depending on the flavor, texture and age of the product to be sold. The U.S. Government sets the exact stan-

dards for rating or scoring the butter product. Grade AA butter has the highest attributed score of 93. Grade A has a score of 92. Grades B and C have ratings of 90 and 89, respectively.

There are five basic kinds of butter products that come in an array of packaging styles. The five types of butter are: sweet butter, sweet cream butter, whipped butter, cultured butter and clarified butter or ghee.

Sweet butter is unsalted butter made from fresh sweet cream. Because it is sodium free, its flavor is delicately mild and smooth. It is for this reason the sweet butter is the spread of choice for fine pastries and French delicacies. We always start with sweet butter when making ghee.

Sweet cream butter is also churned from fresh sweet cream, but has approximately 3% salt added. This is the table butter that we are most used to eating. In fact, most Americans associate butter with that distinctive salty taste.

Whipped butter refers to either sweet butter or sweet cream butter that has been pumped with air or inert gases to create a light, fluffy texture. It's favored for its spreadability, is less concentrated than butter and contains less fat than regular butter.

Cultured butter is also called sour cream butter because lactic acid friendly bacteria are added to the cream. It is then set aside to ripen. This process enlivens the butter with a distinctive, almost tart, flavor. It does help to preserve the shelf life of the butter and is more commonly sold in Europe than in the United States. Once, by accident we received cultured butter instead of regular unsalted butter for our ghee making. Even after completing all the steps of boiling and rarefying to make the cultured butter into ghee, it still had the yogurt-like scent and ever-so-slightly sour taste.

Ghee, or clarified butter as it is more commonly called, is traditionally made by boiling unsalted butter at specific temperatures until the water content of the butter is boiled away and the curd has turned into a sediment that can be discarded. Ghee is the only form of butter that is shelf stable—that is, it does not need refrigeration and will not rancidify at room temperature.

He causeth the grass to grow for the cattle,
and herb for the service of man.

—Psalms 104:14

BETTER THAN BUTTER
Definition of Ghee

Ghee: (GE) Noun. In India, the liquid butter remaining when butter from cow or buffalo milk is melted, boiled and strained. [Webster's New World Dictionary.]

Ghee is pronounced like this: With a hard "G" as in "GO," the "H" is silent and the two "E"s are long. I have received many calls over the years from people who have just purchased their first jar of ghee. They'd say to me, "I'd like to know how to cook with glee." I was always amused that people called it glee and since that word describes the product so well, I never felt the need to correct them.

Ghee is called The Royal Oil *because it was the oil of choice for the wealthiest class in India. Cooking with this liquid gold was a luxury of the aristocracy. Today, of course, ghee and good quality butter are available at most specialty food stores, so the connoisseur* can easily make it a part of their everyday cooking.

Ghee is commonly referred to as clarified or drawn butter, but there are implied differences in the way each of these are processed.

In clarified and drawn butter, the butter is melted and the top layer of solids is skimmed off. It is then considered ready to serve. This makes for a nice presentation on the dinner plate, but it is not ghee. Butter clarified in this way still contains most of the milk fat solids and the water content that is inherent in the butter. This will result in the same freshness breakdown that can be seen in butter: spoilage, rancidity and mold after a few days without refrigeration.

When ghee is made using the traditional boiling and straining method (see the chapter on making your own ghee), all the water is boiled away and the milk fat solids are completely removed. The result is a glistening clear, golden oil that will not go rancid. This is the liquid gold that is described in ancient Indian scriptures. And this is the ghee that will find its way into all of your cooking.

In addition, properly made ghee will have the following qualities:

- it is salt free
- it is lactose free
- will not burn or smoke at high temperatures
- it needs no refrigeration

For sheer culinary perfection, ghee has long been the choice of Europe's great chefs. For medicinal purposes, ghee has been used by Natropaths in many cultures, drawing on its curative and rejuvenating powers. And for those with severe dairy allergies, ghee is an integral part of their often highly restricted diet. In each application, ghee serves the purpose of enhancement.

In the ancient scriptures of India, the word ghee is sometimes used interchangeably with the word life. To me, that is the most definitive way of describing this rarefied oil of butter: Ghee is Life.

And it will come to pass in that day… the hills shall flow with milk.

—*Book of Joel 3:18*

HIGH TECH vs. HIGH STANDARDS
Organic vs. Conventional Butter

Certified Organic butter is available in health food stores and specialty markets nationwide. (See the organic dairy products listing in the back of this book). When you purchase organic products, you are casting a powerful economic vote for sustainable agriculture. Also, organic methods encompass the humane treatment of animals and environmental balance for our future generations.

I plead total ignorance to the workings of farms and agriculture in general. I grew up in the suburbs. My first exposure to real cows was when I was eight years old and our Brownie Troop went on a field trip to a dairy farm. I remember concluding that it must be wonderful to be a cow...expansive green pastures and long, lazy afternoons sitting in a grassy field. At one point in our tour, the farmer winked at us and said, "And of course you know that the brown cows give the chocolate milk."

"Oh my," I thought, "I didn't know that."

I looked at the brown cows with new admiration and for weeks I was wondering what kind of milk the spotted cows gave.

In the world of today's agri-business and cow factories, as they are called, it's a very different story. My childhood memories of dairy farms and pristine milk cows make the realities of conventional dairy farming even more shocking.

Our bovine friends, whose natural life span was about 13 years, now have a life expectancy of around five years. This discrepancy is a result of modern farm techniques, depleted living and grazing conditions, over breeding and the inundation of drugs. Today's milk supply is so embedded with growth hormones, pesticide residues and drugs that, in the words of Washington Post *columnist, Colman McCarthy, "It should be sold by prescription only." A 1989 investigative report by the* Wall Street Journal *quotes an independent agency that found drug residues in 40% of the milk samples taken from ten different American cities.*

The Food and Drug Administration does random testing on the milk supply, but the testing itself is done on a range of 12 different drugs. The problems lay in the fact that there are over 60 different drugs that are administered to the cows! In a 1992 government report, the Government Accounting Office states:

> *FDA inspection data from 1990 and 1991 indicate that 62 animal drugs not approved for use on dairy cows, and 42 drugs not approved for any food-producing animal, were found on dairy farms across the nation.*

The latest and most controversial drug being given to dairy cows is the Bovine Growth Hormone (BGH). This a genetically engineered synthetic hormone that, when given to a cow, increases her milk output by about 10%. This hormone is identical to a hormone that the cow produces naturally and at this time, there is no test to detect the BGH in the milk supply. In the last months of 1993 the Food and Drug Administration ruled BGH safe to use on dairy cows. The Codex Alimentarius, an international standards organization in Italy, came to a different conclusion. In meetings held in the summer of 1995, the 14 nation European conference opposed the use of synthetic BGH. The European community of nations as well as Canada have placed a moratorium on the drug.

With the use of BGH there is an increased risk of disease for the cow. The manufacturer of the drug, in its brochure for veterinar- ians acknowledged 21 different side effects of BGH, including an increase in mastitis. The Consumer's Union has filed data showing a 79% increase in mastitis with the use of BGH on dairy cows. Many new drugs must be given to the cows to combat the mastitis and other diseases brought about by the use of synthetic BGH.

This is not to say that all dairies are bowing to the use of BGH and other genetically engineered supplements. A growing number of dairies, large and small, are requiring that their contributing farmers sign agreements on the abstinence of these drugs.

Now, let's take a look at the Cream of the Crop—the organic dairy farms. These are generally small family-owned farms that have 60 to 80 cows. They are usually part of co-operatives who pool *their dairy products so they are able to move into the marketplace with all the clout of a larger company. These dairy farmers adhere to stringent standards of purity and environmental responsibility. The Organic Crop Improvement Association is one of the leading organic certification guilds. Some of their requirements for organic certification include: The farmers cannot use feed crops grown with the aid of pesticides, herbicides or synthetic fertilizers. They must not inject any synthetic hormones or drugs into their cows and if a cow should become sick it is taken out of milk production for twice the time specified in government regulations. In addition, there is a code of ethical and humane treatment in the handling of the animals. This includes such basic quality of life considerations as allowing newborn calves to be nursed by their mothers, and not allowing calves to be sold to veal processors. The cow that grows up on an organic farm is assured free-range grazing and a life more attuned to nature's plan.*

In 1990, I met Harriet Behar from Organic Valley, the leading organic dairy co-operative in the country. She told me they were going to start producing certified organic butter and that this would be a great opportunity for Purity Farms, Inc., to start making organic ghee. At the time, we were using a high quality kosher butter from small family farms. Now that certified organic butter was becoming available to us, there was one major consideration: THE COST. Organic butter was going to cost us 150% more! Howard and I looked at each other and said, "Well, if we are going to be in business, we want to only sell the very best—no matter what. There have to be a few people, just like us, who buy for quality, not just price." What appeared to be a risk in our move to organic butter, turned out to be the greatest boon for our company. Looking back, it was all just a matter of commitment. So honor your ghee-making ritual with a commitment to using the best possible butter available to you. It's a small choice with global ramifications.

If you're trying to eat less fat, spray or barely drizzle
warm ghee on plain steamed or baked vegetables.
A few drops of ghee and freshly chopped herbs adds
a wonderful garnish to unadorned fat-free soup,
rice, pasta or cooked beans.
—Yamuna Devi, Author

After the ice cream, of course, comes the coffee, served Demi-
tasse.
That's to prevent any possibility of drinking out of the saucer.
—Will Rogers, American Humorist

You are become such as have the need of milk, and not of strong
meat.
—Book of Hebrews 5:12

UNCLOGGING THE PIPES
Cholesterol and Ghee

Say the word cholesterol to people and it conjures up negative mental pictures of clogged arteries and other degenerative diseases. Certainly this is part of the story, but the role of cholesterol in our health has been over generalized and over simplified in order create a sound bite approach to wellness. Natropath John Finnegan has written several books on the quality of fats and cholesterol in our diets. A quote from one of his recent papers reads:

> Cholesterol is the precursor from which all our vital adrenal and sex hormones are produced. Cholesterol is also a potent antioxidant that floods into the blood when we ingest too many harmful free-radicals, found in damaged and rancid fats from margarine and highly processed vegetable oils. It promotes the health

of the intestinal wall and protects against cancer of the colon and plays an important role in the development of the brain and the nervous system.

Cholesterol in its pure form is a light colored powdery substance found in all cells of animal tissues. It is not water soluble, so the body creates lipoprotein bundles to carry the cholesterol through the blood stream. These little bundles of proteins that carry the cholesterol can either be HDL (High Density Lipoproteins) or LDL (Low Density Lipoproteins). High levels of LDL are associated with coronary heart disease while HDL are believed to help keep the arteries clean by assisting in the removal of the LDL cholesterol from the linings of the arterial walls. Our propensity of having high HDL or LDL levels in our bloodstreams is partly hereditary, partly diet and life-style enhanced and is also influenced by liver function, which produces cholesterol for our hormones, tissues and digestive acids.

Until recently, the focus has been on cholesterol levels only. Now researchers, such as John Finnegan, are looking closely at the role that trans-fat plays in arterial dysfunction. Trans-fatty acids are toxic fats in oils and margarine resulting from modern methods of hydrogenation and refining. In a recent article, Finnegan writes:

> *Look at it this way, at the turn of the century people in North America ate six times the amount of butter that we do today. Yet now on a consumption of butter, one-sixth as previous and yet hundreds of times higher in trans-fats, we have an astronomical amount of obese (excuse me, slightly overweight) citizens. Not to men-*

tion that one hundred years ago there was virtually no heart disease (now it strikes two out of three), cancer was extremely low (now one out of three), and so on.

So how does ghee fit into this picture? In India, ghee has been used for thousands of years as a basic cooking oil. It is made from butter, which is an animal product, so therefore it contains cholesterol and saturated fat. So how then, does ghee figure into the diet of a person who is conscientious about fat and cholesterol intake?

To answer this, I want to refer to a study that I read in Finnegan's popular book, The Facts About Fats. *In it, he reprints a portion of an article that first appeared in the* American Journal of Clinical Nutrition, 1967, 20:462-75. *It follows:*

> *The study was performed by Dr. Malhotra, a medical doctor for the India National Rail System. He found two population groups in India, one in the North and one in the South. The Northerners were meat eaters, and the main source of fat in their diets was ghee (clarified butter). You might assume, therefore, that they had high cholesterol levels. You would be right.*
>
> *The Southerners were vegetarians, with much lower cholesterol levels. Even so, they had 15 times the rate of heart disease compared to their northern neighbors. The major dietary difference Dr. Malhotra found was in the kind of fat the Southerners used. They had abandoned the traditional use of ghee-real food-in favor of plastic-food margarine and refined polyunsaturated vegetable oils.*

In any given day at Purity Farms, Inc., I receive interesting phone calls from people who share their experiences with ghee or ask specific questions about using ghee. One of the calls I received was from Dr. Narinder Saini, at the Cardiac Rehabilitation Center in

Springfield, Ohio. He had just completed a book called Create a Healthy Heart: I Did It Why Don't You? *It is an account of reversing his own coronary heart disease with diet and life-style changes. Dr. Saini is aware of the role that ghee plays in his own heart-healthy diet. He suggested that I get a trans-fatty acid test done on our ghee. He felt that ghee must be very low in trans-fats which accounts for the long-held belief in India that ghee promotes health, including cardiac health.*

On his insistence, I had our ghee tested and, sure enough, the trans-fatty acid count was so low that the chemist said our ghee could be called quantitatively free of trans-fat.

Dr. Saini graciously wrote the following paragraphs for me to include in this book:

> *Ghee or clarified butter is a better choice than margarine and refined oil. Ghee has been used in Ayurvedic prescriptions and it is the way of cooking in many cultures since about 2,000 years ago. It has a unique healing property not found in other oils. According to Ayurvedic teachings, it enhances the oja's, an essence that governs the body and balances the essential body hormones. To ensure a sound body and mind you must have more oja's. It also increases the longevity and has anti-viral and anti-cancer properties.*
> *Cholesterol itself is not harmful but becomes harmful when it is partly broken down or oxidized by unstable chemicals called free radicals. Ghee and particularly organic ghee has no oxidized cholesterol or trans-fatty acid and is very stable at higher cooking temperatures. Because it is stable and does not become rancid it can be stored for long time.*
> *Ghee, because of its biochemical properties, will use*

30% to 40% less amount of oil compared to any other oil. One tablespoon of oil or ghee has 14 grams of fat but for cooking you will use about 40% less ghee so it means for the same cooking you will add only 8.4 grams of fat.

It is also known that it has a very strong anti-viral properties. If you have allergies then dip your index finger in ghee and sniff through your nostril, it will make a protective coating on the mucosa so that the allergens cannot create the allergic reactions.

So many other properties are still under investigation and will soon be made public.

One must ask children and birds how cherries and strawberries taste.

—*Johann Wolfgang Von Goethe, German Poet and Philosopher*

Examination of Ghee
Complete Nutritional Analysis

	per 100 gms	per 5 gms (one serving)
Calories	895.00	44.77
Moisture gms	0.29	0.01
Ash gms	0.04	0.00
Protein gms	0.33	0.02
Carbohydrate gms	0.00	0.00
Fat gms	99.34	4.97
Sat-fat gms	61.10	3.05
Mono unsat-fat gms	28.50	1.43
Poly unsat-fat gms	3.90	0.19
Cholesterol mg	165.60	8.28
Fiber gms	0.00	0.00
Vitamin A RE	907.00	45.35
Vitamin C mg	0.00	0.00
Sodium mg	1.48	0.07
Calcium mg	4.23	0.21
Iron mg	0.12	0.01

Note: None of the fat appears to be hydrogenated
Reprinted by permission of Purity Farms, Inc.

Condensed milk is wonderful.
I don't see how they can get a cow to
sit down on those little cans.
—Fred Allen, American Humorist

Reacting To The Reactions

Dairy Allergies and Ghee

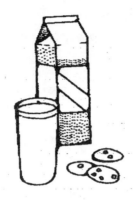

An allergic reaction is the body's response to a perceived foreign invader. In the case of food allergies, lactose intolerance is surprisingly common among adults over 20. A person who is lactose intolerant will often refer to their 'dairy allergy'. Specifically, that person produces inadequate amounts of lactase in the small intestine to digest the natural sugars in dairy products. It is rarely considered life-threatening, but it can be chronic and severely limit the foods in a person's diet. For the person with lactose intolerance a bowl of ice cream or a glass of milk can result in a speedy reaction ranging from bloating, diarrhea, sinus irritation or hives.. Although butter has much less lactose than milk or cheese, some people cannot have even <u>one</u> pat of butter on their foood without consequences. Ghee, being lactose free, can be a welcome addition to an otherwise highly restricted diet. This also holds true for those with candida (yeast organism imbalance) or immune system disorders. Check with your health practitioner about including ghee in your diet if you have lactose intolerance.

In the process of boiling butter to make ghee, the milk fat particles, which contain lactose or milk sugars, cling to the bottom of the pan as a lightly browned crust. These cooked on milk fat solids are discarded as the pure oil of butter (ghee) is poured off and sealed in jars. So, traditional ghee, is by nature, free of lactose and proteins. I've always thought of ghee as a non-dairy product because it doesn't share many of the same properties with milk or cheese. It doesn't need refrigeration, it doesn't go rancid and it is actually good for people who cannot digest dairy products.

In the source section in the back of this book, I've listed a few comprehensive resource books for people with food allergies.

As Dr. David Simon, Medical Director of The Chopra Center for Well Being states:

> According to Ayurveda, ghee has the remarkable distinction of enhancing well-being in almost every conceivable circumstance. It nourishes, detoxifies and willingly carries healing herbs to our cells and tissues. It is no wonder that ghee is seen as a gift from the gods.

Borscht and bread make your cheeks red.
                              - Jewish Folk Saying

I saw him even now going the way of all flesh, that is to say towards the kitchen.
                              -John Webster, English Dramatist

NATURE IN BALANCE
Ayurveda

My first encounter with ghee came through the study of Ayurveda. Ayurveda is a Sanskrit word that translates Science of Life. Ayurveda is the branch of applied literature that deals with health and healing in Vedic scriptures. Based on the Samkhya philosophy of knowing truth, the ancient Rishis recognized the principles and practical applications of this life science. For the past 5,000 years, Ayurveda has been an inherent part of daily life in India.

This precise and multi-faceted approach to health is based on man's energetic link with the cosmos and the way the five elements, earth, air, fire, water and ether, weave through every animate and inanimate object. Based on this eternal wisdom, Ayurvedic scriptures cover every aspect of life and health from diagnostic procedures to healing remedies to development of consciousness and longevity.

In every way, Ayurveda is a fresh, dynamic science of discovery that has some enlightening knowledge for every nuance of daily living. In the back of this book I have listed several excellent books on the science of Ayurveda, as well as health centers that specialize in the practice of the Ayurvedic medicine.

As an Ayurvedic doctor once told me, "Ghee is Life. You people in the West have such a high rate of senility because you take in the wrong oils all your life. If you would just have a little ghee each day, you'd be sharp and wise into your old age."

The Charaka Samhita, *the Ayurvedic Text on Health*, describes ghee this way:

> It is promotive of memory, intelligence, vital fire, semen, vital existence (ojas), Earth (kapha) and fat. It is curative of vata (air), pitta (fire) and toxins.

Ghee is one of the primary digestive aids used in Ayurveda because it stimulates the secretion of the stomach acids needed to properly digest food. As a prescription for constipation, ghee is taken in warm milk. Besides being regarded as a general tonic for healthy eyes and skin, ghee is used to allay the symptoms associated with colitis and peptic ulcers. Most notably, the ghee is a carrier of medicinals to the tissues of the body. Throughout the Ayurvedic texts are formulas for herbal remedies and rejuvenation decoctions that are mixed in a base of ghee. The deep, penetrating quality of the ghee acts as an agent of fulfillment—taking the intelligence of the herbs to the affected cell in order for the medicinal to do its healing work. It is said in the Ayurvuedic texts that as ghee ages, its medicinal properties increase. In many Indian households ghee is aged to hand down to the next generation. In India there is a saying, "One hundred year old ghee can cure anything"

In regard to the skin, I can personally attest to the healing power of ghee on burns. We've all heard the old folk remedy of putting butter on a burn. That will, in fact, create a blister. But if ghee is put on a burn it will most likely heal without a scar. Over the years I have had hundreds of tiny burns from the bubbling over of ghee and yet, there has never been a blister or a scar. I have heard many stories of people who were accidently burned and quickly reached for the ghee jar and covered the affected area with ghee.. The results were always no blister and no scars.

For cosmetic or medicinal purposes, the references to the curative properties of ghee in Ayurvedic medicine are rich and diverse. In addtion, each text on the healing sciences in Ayurveda is voluminous and incredibly detailed. Attaining mastery in any area of Ayurvedic medicine could be the worthy study of a lifetime. To that end, I include this excerpt on ghee from Swami Sada Shiva Tirtha's reference encyclopedia on Ayurveda:

Ghee (ghrta) Energetics: Sweet/cold/sweet
Actions: Tonic, emollient, rejuvenative, antacid
Indications: Fattening, increases marrow, semen
and ojas. Improves intelligence, vision, voice,
liver, kidneys, and brain.
The best form of fat for the body. It also

satisfies the craving for meat. It is the best oil for pitta dosha. All the digestive fires (agnis) are balanced from eating ghee. It is good for memory and digestion, useful in conditions of insanity and consumption, promotes longevity and reproductive fluid, good for children and the elderly. It develops and maintains suppleness of the body and lungs, is useful for herpes, injury, vayu and pitta disorders, fevers, TB and is highly auspicious and holy (sattvic). Taken with herbs, it transports their nutrients and energies to all seven tissue layers. When special herbs like ashwagandha are made into a medicated ghee they remove harmful cholesterol from the body. Many different herbs are made into medicated ghees for enhanced healing effects.

[Taken from the Complete Ayurveda; A Reference Encyclopedia for Families and Practitioners by Swami Sada Shiva Tirtha (unpublished). Reprinted with permission.]

The purified soma juices have flowed forth, mixing with curd and milk. They are cleansed in the waters.

—*Rig Veda IX 1.24*

It is promotive of memory, intelligence, vital fire, semen, vital essence (ojas), kapha and fat. It is curative of vata, pitta, fever and toxins.

—*Charaka, Book of the Vedas*

WHY COOKING IS SUCH A JOY!
Rituals of the Kitchen

If a person is inclined to look at life in a spiritual context, cooking then, is a ritual of great importance and value. The Vedic texts, which extol the medicinal and spiritual values of ghee, go into great detail about behavioral Rasayanas. A Rasayana is something that promotes longevity and rejuvenation. So the behavioral rasayanas instruct a person into the ways of gracious and sattvic (life enhancing) living. Although there are rasayanas that relate to every aspect of life from relationships to herbal supplements, I would like to list some of the guidelines that apply to your time in the kitchen making ghee or preparing any other foods.

"Cleanliness is next to Godliness." This certainly applies to the kitchen. One of the first considerations of cooking space is that it be sparkling clean and free of excess clutter. Preparation for cooking begins before you enter the kitchen. Ancient scriptures state the importance of washing face, hands and feet and wearing fresh clean clothes into the kitchen. Nails are to be cleaned and trimmed and hair is to be pulled back. Before entering the kitchen, the cook pauses and says a prayer of invocation to create the proper atmosphere—one that results in healthy dishes, nourishing all aspects of the cook and the recipients of the food. Negative emotions are left at the door, so to speak. In addition, women observe the rule of not preparing food during their menstrual periods.

Once in the kitchen, street shoes are not worn and jewelry is removed. Cooking is done in silence or with gentle, soothing background music. Various mantras are used while cooking. Man-

tras are sacred sounds that have the capacity to settle the mind and refine the thought process. Out of reverence for the oral tradition in which these are passed down and taught, I do not list them here. But, I do offer you a simple prayer I have used for years and years, since I began making ghee: Lord bless this Ghee. May it give Light and Life to all who use it.

After a good dinner, one can forgive anybody, even one's own relations.
—Oscar Wilde, Author and Critic.
A crust eaten in peace is better than a banquet partaken in anxiety.
—Aesop, Greek Philosopher

STARTING FROM SCRATCH
Making Your Own Butter

Churning your own butter from fresh whipping cream is a simple, straightforward process. Most of us have inadvertently started that process when we over whipped the whipping cream for strawberry shortcake.

Start with certified organic whipping cream (minimum 30% fat content). Allow the cream to sit at room temperature for 20 minutes. The temperature of the cream will affect the finished product. Churning butter on a warm sunny day will result in a butter with a soft consistency and pearl-like color. Whereas, churning the cream at a cooler temperature on a cold winter day will create a butter with a waxy, stiffer texture. The ratio to keep in mind is that a full gallon of fresh cream will yield a little over three pounds of butter. Besides hot and cold weather, the humidity will also affect the churning process. If you live in a humid climate or it happens to be a rainy day, allow up to 20% longer for the cream to be churned into butter.

Fill a clean stainless steel, crockery or glass bowl about half

*full with cream. You may also use a variable speed food processor.
If you are using the hand method, begin agitating the cream with an
old fashioned egg beater or electric mixer on low speed. By hand, the
churning process takes about half an hour; by processor or electric
mixer, it takes about five minutes. For those people with Herculean
arm strength and patience, you can also churn the butter by putting
the cream in a jar with a tight fitting lid and begin rotating the jar
in an infinity (figure eight) motion.*

*As you begin the churning, the cream will be foamy and
resemble thick dish-soap. Next, it will form into a mushy oatmeal-
like mixture. Then, as the butter is about to separate, you will see
the bead-like fat particles forming. Scrape the bowl for all the butter
pearls and knead them into the form of a ball. With a wooden spoon
or spatula, pour off the remaining buttermilk into a container and
securely cap and refrigerate it for future cooking needs.*

*Wash the butter ball under cool (not hot) running tap water. Pat
dry with paper towels.*

*Place the butter back into the mixing bowl and, if you wish to
have salted butter, take the wooden spoon or mixing paddle and fold
in the equivalent of 1/2 teaspoon of salt for every pound of butter.
Press the butter into a container with a tight-fitting lid or wrap it
securely in plastic wrap.*

*Your freshly churned butter will be a very pale whitish yellow.
The scent will be delicate and creamy. Use the butter within 30 days*

if storing it in the refrigerator or before 6 months if you are freezing it at 0ºF or -18ºC.

If you are going to use the homemade butter to make ghee, you can use it within either of these time frames. Use only unsalted butter for ghee making.

Cheese—milk's leap toward immortality.
　　　　—Clifton Fadiman, American Author and Critic

YOU HAVE EVERYTHING YOU NEED
Utensils for Ghee-making

In your kitchen, you probably have everything you need to make a batch of ghee. Purchase two pounds of good quality (preferably organic) butter. In the back of the this book, I have the address and phone number of the Pure Food Campaign and they have lists of BGH free and certified organic brands of butter.

Use a heavy stainless steel four quart pot. Aluminum is an unacceptable metal to be used for ghee making, as is copper and cast iron. I have found glass sauce pans to have uneven heating distribution, although I love doing every day cooking in my glass cookware. Making ghee in copper kettles results in a chemical reaction that may affect the oxidized cholesterol levels in the ghee. Copper-clad bottoms on stainless steel pans are fine. They really do encourage even heat distribution, so they are ideal. Scrub the pot with soap and water and dry thoroughly.

A long stainless steel slotted spoon will be good for stirring. Please do not use wooden spoons, as they can hold moisture and bacteria. Dish cloths, pot holders, and an apron will come in handy. Although ghee won't burn your skin, it will make a stubborn oil stain on your clothes, so the apron is highly recommended!

A cookie sheet, cutting board or drain board should be on hand to set your jars on while pouring the finished ghee. In addition, a wide-mouthed strainer is handy for many kitchen tasks, (see the source guide in the back of this book). Also keep a piece of cheesecloth handy and some paper towels within reach in case of spills.

Do you have a fire extinguisher for your kitchen? For stove-top cooking with oil, it is a good idea.

You can use any variety of jars for your ghee. Make sure they are glass, not plastic. Ghee will absorb that toxic plastic odor and the plastic cannot take the high temperatures of hot ghee. You can recycle jars, but avoid pickle, salsa or mustard jars. Not only do the caps require special sealing devices to make them air tight, but the caps hold the smell of the food that was formerly in the jar. An old nut butter jar with a screw on cap will work fine. My all time favorite jars are Weck canning jars from Germany. These containers come in a few small and medium sizes, with nice glass tops (see the 'sources' in the back of the book).

To sterilize the glass, dip the clean jars into boiling water for a few minutes and take them out with a pair of tongs. Running jars through the dishwasher is also an adequate way to clean them for your household use. The key is to make sure the jars and lids are completely dry before pouring the hot ghee into them.

"I want a round table for a square meal"

> *-Percy Feldenkreis, Furniture Maker*

"Plain cooking cannot be entrusted to plain cooks"

> *-Countess Marcelle Morphy, Cookbook Author*

MINING YOUR OWN LIQUID GOLD
Making Your Own Ghee

Ghee has a long history that spans many cultures, so it is understandable that different methods of making ghee have evolved. The most straightforward way is to use the stove-top method. This affords you the most control and ease in regulating the temperature. When cooking with hot oil, both of these factors are important.

Commercially, there is a clarified butter on the market that is made by spinning the butter using centrifugal force to expel the water and create a butter oil product. This, of course is not feasible to do in a household kitchen! Although this method does result in an oil that will have some of the same qualities as ghee, it has a petroleum jelly-like texture and a distinctively unpleasant scent.

Ghee can also be made in the oven by melting the butter in a large flat bottomed casserole pan. Some people swear by this method of ghee-making, although it has varying degrees of results.

The oven method creates a possible fire hazard, since the ghee can splatter while it reaches the boiling stage. In addition, the ghee is hard to check for doneness and once the butter has turned into ghee, you have to negotiate this wide pan of hot oil into these little jars. Again, the stove-top method avoids all these problems and, overall, will take less time than the oven method.

The traditional stove-top method of making ghee follows:

In preparing your ghee you don't need thermometers, clocks, or any other instruments. Your sense of sight and smell will be your guide as you turn butter into ghee. There are many variables such as altitude or humidity that will alter the time that it takes to make a batch of ghee. In addition, if you make a large batch of ghee (8 to 10 pounds), expect to spend more time in the kitchen while it cooks. Generally, it will take under an hour to make a two-pound batch of ghee from start to finish.

- *First, cut up two pounds of butter into one inch squares and place in a four quart sauce pan.*

- *Turn the heat to medium and let the butter melt slowly. Depending on the efficiency of your stove-top you may want to turn the heat up a little until the butter comes to a boil. You want to avoid scorching on the bottom of the pan, so be cautious with the high heat settings.*

- *With the heat at the low to medium range, let the*

butter continue to boil. The ideal is to maintain an even, rolling boil, bubbling over most of the surface of the mixture. It will be a milky-whitish color.

- *Often as it begins to boil, the butter will start frothing and rising toward the top of the pot. If left to its own, it would boil over. This stage of frothing lasts no more than five minutes and is the result of the water content of the butter rising to the surface to get boiled away. Do not scrape off this frothing foam. Instead, take your long slotted spoon and gently stir the ghee from the center of the pot. I always keep the stirring restricted to the top one-third of the pot and I make very small clockwise circles, about the size of a silver dollar. This simple step allows the moisture to have a channel out. Almost immediately, you will see the foaming settle down and the ghee will continue to boil. It will be a brighter yellow now and the scent of butter will start to fill the kitchen. Double check to see that the heat is not too high. If it is too hot, little burnt particles will be floating to the top.*

- *Now just let the boiling continue, and nature will do the rest! The gentle, constant boiling action is separating the milk fat particles from the oil and allowing the moisture to rise to the surface and evaporate away in the steam. Taking into account the variables of humidity, quality of butter, etc., this part of the process can take up to 25 minutes. You don't have to stand over the ghee the whole time, but do check in on it every few minutes and give it a stir now and then. The steam will continue to rise and the color will be a brighter canary-yellow and the scent will be sweet.*

- *Now, begin watching the pot closely. The butter is turning into ghee. The yellow liquid is appearing more golden and clearer and you can actually see the bottom of the pan. At the bottom, is a layer of sediment that is golden to coppery brown. You will also see a bit of crustiness around the edges of the pan. The aroma will have changed from sweet to nutty, not burnt, but rich and roasty.*

- *When the boiling liquid is hot bubbly oil and there is barely a trace of steam coming from the surface, you have ghee—pure, clarified oil of butter, liquid gold, as it has been called for thousands of years.*

- *Turn off the stove and place the pot on an unheated burner while you prepare to dispense the ghee.*

- *Line up your clean, dry containers on a tray, board or countertop that is impervious to heat. Make sure the containers are at room temperature, since hot oil can break cold glass.*

- *Place the piece of cheesecloth in the hand held strainer and balance it on the top of the jar if your pot full of hot ghee does not have a handle arm. By this time, the ghee has had a chance to simmer down for a few minutes and any floating bits of sediment have settled to the bottom. Pour the ghee through the cheesecloth strainer into the first jar and then set the pot back on the stove top and cap the jar immediately. The hot ghee will conduct heat to the glass very quickly and if you wait even a few minutes the glass will be too hot to handle. Continue to pour the ghee into each jar, stopping to cap each one.*

- *Let the capped ghee sit while you clean up the pot. Some people save the burned solids from the bottom of the pan and use them for frying, but, to me, that never seemed to make much sense. If I wanted those cholesterol-laden fat particles, I wouldn't bother making the butter into ghee in the first place. I always discard the sediment from the bottom of the pot.*

- *Touch the jars of ghee and check the temperature. If the glass is warm, not hot, you are ready for the final step of the process. Howard and I developed this final step to give the ghee a smooth consistent texture and color. We noticed that many home-made ghees looked thick at the bottom and watery at the top—the color ranging from light to dark yellow in the jar. They would also have a grainy texture that was all-around unappealing. To create that perfect jar of ghee, take the almost room temperature jars of ghee and put them in the refrigerator until they are jelled. It takes about 40 minutes in the refrigerator to get that uniform, opaque color. Even in warm weather, the ghee will retain its smooth texture and consistent color at room temperature if it has been jelled. Some people like to put the ghee in the freezer to solidify, but I caution you not to do that for two reasons: The abrupt temperature change can cause the glass jar to break, and very quick cooling of the ghee causes crystallization to form at the top of the jar.*

• *Store your ghee in a cool dry place or keep the jar you are using at hand and refrigerate the rest. It will store well at room temperature and is shelf stable. Properly made ghee will last for many months. In theory, ghee does not spoil because it has no water content and no milk fat solids to go rancid. It will change color over time though. There may appear to* be a waxy whitish gloss to very old ghee and it may take on a crystalline texture if exposed to very cold temperatures. If your ghee is subjected to very hot or cold temperatures, resulting in either a crystalline texture or stratification of the ghee you can rebalance it in this way: Simply take off the lid of the jar of ghee and place the jar in a saucepan filled one-third full with warm water. Turn the burner on low to medium low and let the ghee melt in the hot water bath. Make sure that no water gets into the ghee jar. When the ghee is a clear liquid again, remove it from the saucepan, dry off the jar and replace the lid, holding the jar with a dishcloth or pot holder, if necessary. Now place the jar in the refrigerator to let it jell up to an opaque light yellow color. Now the texture will be smooth and the color consistent from top to bottom. In each of these cases, the ghee is still fine to use and will taste the same in all cooking and baking applications. You can freeze ghee, as well. I don't do that, since I always have a lot of it around, but it is an option for safe, efficient storage.

- As for the issue of mold, ghee can grow mold if water is introduced into the jar. That can be done as innocently as using a spoon with food particles on it to dip into the jar, or storing the jar, opened and in a humid environment. When my husband and I lived in Washington, D.C., I was amazed at the air-borne particles that settled into a jar of ghee that I accidentally left out. So simple, common sense tells you to always use a clean utensil to dip into the jar and keep the lid of the jar on when not in use. Leaving your ghee at room temperature, in a cupboard or right on the counter is the recommended way to store it. If water does get mixed in with the ghee and mold grows, simply scoop out the tainted part and the rest of the ghee is still edible.

When the ghee is done, we usually pop a few pieces of bread into the toaster and then top the toast with the freshly made ghee— it is divine!

You have just made something very special!
Enjoy your ghee in good health!!!

The best way to learn to cook is to cook; stand yourself in front of the stove and start right in.
—Julie Dannenbaum, American Cooking Teacher

HOLY SMOKE!
Recycling Burnt Ghee

If you should accidentally over cook your ghee, don't worry—it has happened to all of us and there are a few options: First, if you've over cooked it a little, the color will be a darker, duller yellow. The aroma will be very nutty and will be devoid of the sweetness. In this case, you can mix it with salt, spices and herbs and use it over vegetables or rice. You don't want to expose the ghee to any more heat via cooking or baking.

If the ghee is severely overdone it will appear a dark coppery yellow and it will have a charred smell. At this point, the ghee has no value to you in the kitchen—but the birds will love it!

My husband, Howard, had this creative idea for one of my unsuccessful batches of ghee. He took the pot with the burnt ghee and stirred into it sufficient black sunflower seeds and millet to make a stiff, lumpy mixture. Next, he put the whole pot in the freezer overnight and in the morning he pried the ghee/seed mixture out of the pot, put it on a piece of waxed paper and took it out to the bird feeder. Of course, this little concoction is better for winter-time bird feeding, but it certainly was eaten up fast by all of our feathered friends. Think of this as a very high-class suet!

You can also recycle the burnt ghee into ghee candles (see page 71). Be sure to add some citronella, vanilla or peppermint oil to mask the scent of the slightly burnt ghee.

If this is coffee, then please bring me some tea.
But if this is tea, please bring me some coffee.
            —*Abraham Lincoln, 16th United States President*

YOUR KITCHEN PHARMACOPOEIA
Medicinal and Culinary Ghees

FLAVORED AND MEDICINAL GHEE

Throughout the Ayurvedic texts, herbs and spices were added to ghee for medicinal purposes. That is, the healing essence of the herb is boiled into the ghee. Because of ghee's powerful absorption qualities, it carries the healing power of the herbs into the tissues and organs quickly and efficiently. Edgar Cayce, the great 20th century psychic, also prescribed ghee and herbs for specific ailments in some of his readings.

For culinary use, you can be endlessly creative! Over the years we have been approached to make flavored ghees for seafood restaurants or Indian spice companies. Due to the absorbing quality *inherent in ghee, we found that a spiced ghee boiling in one kettle would make all the ghee in all the kettles have that same scent. So, if you decide on a flavored ghee, make it by the batch and not concurrently with your regular batches of ghee.*

The following recipes will get you started in the art of making exotic ghee potions!

COMMON MEDICINAL GHEES

Licorice Ghee

Heat one part pure licorice powder to eight parts of water. Once the mixture comes to a boil, turn the fire down to medium low and continue to boil the liquid until it is three quarters boiled away. Take the one quarter mixture that is left in the pan and stir in an equal amount of ghee. To this paste mix in an equal amount of water then return to boiling and keep at medium low heat until all the water is boiled away. Transfer the freshly made licorice ghee into a glass container, seal and refrigerate.

To eat is human; to digest is divine.
—*Charles Townsend Copeland, American Educator*

Ginger Ghee

In your ghee-making process, once the butter has melted add one quarter cup peeled, sliced ginger. Very fine, almost transparent slices or minced ginger results in the finest quality ginger ghee. When you strain your ghee at the end of the ghee-making process, you will catch all the ginger fibers in the cheesecloth leaving a clear ghee, with a heavenly spicy and pungent scent.

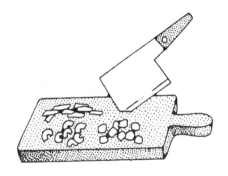

Appetite is the best sauce.

—*French Proverb*

Black Pepper Ghee

Wrap one eighth of a cup of whole peppercorns in cheesecloth and tie securely with cotton cord, not plastic or metal ties. Add the peppercorn bag to the melted butter and let it boil in the pot while the butter is cooking. When you have reached the ghee stage, pour the ghee through the cheesecloth strainer, leaving the peppercorn bag and lightly burnt solids at the bottom of the pan.

There are a variety of peppercorns available in gourmet stores, so you might want to try a combination. My personal favorite remains the whole black peppercorn for black pepper ghee.

Pepper is small in quantity and great in virtue.

—*Plato*

Brown Butter

In France, it is known as Buerre Noisette. *Considered a must for baby asparagus or tender spring broccoli. Simply melt one quarter cup ghee in a saucepan and simmer at low heat until the ghee turns a coppery brown color. Remove from the burner, pour over the freshly steamed vegetables and sprinkle with capers.*

... mankind is divisible into two great classes: hosts and guests.
—Sir Max BeerBohm, English essayist and Critic

FLAVORING GHEE WITH ESSENTIAL OILS

Essential oils are readily available in health food stores and good bakery supply houses. The key is to make sure you get pure oils—not flavorings that contain alcohol and water. Let your own taste buds be the guide in adding oils to the ghee. As a general rule, I start with ten drops of food grade essential oil for every one cup of ghee. Then add two drops at a time until the desired aroma is achieved. Try a few of the following combinations and WOW everyone at your next dinner party!

Lemon Ghee

Add 15 drops of pure lemon oil to one cup of ghee. Serve this with asparagus, broccoli or corn on the cob. This combination seems to stand on its own beautifully without any other spices. It is wonderful for people on a sodium-restricted diet. You can bottle the lemon ghee and place it in a basket filled with baby artichokes—this makes a perfect housewarming gift or holiday basket.

Of soup and love, the first is best.
—Spanish Proverb

Orange Ghee

Start with 10 drops of the orange oil slowly, one drop at a time, since the orange oil can be powerful. This is great on hot cinnamon buns or French toast.

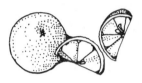

Curried Ghee

This is a must for every well appointed kitchen! It is so easy too. The only basic guideline throughout is to use spices or herbs that are free of any moisture content. You can add ready made curry power of Garam Masala *(a mixture of ground and dry roasted spices) from an Indian food store. Or, you can make a version of homemade curry powder by combining: 1 teaspoon ground cumin, 1/2 teaspoon each ginger, chili powder, cardamom, coriander, turmeric, nutmeg, and cinnamon. Add this mixture to the ghee and add salt and black pepper to taste. Store the ghee in a sealed container and refrigerate. Use this curry paste to sauté vegetables, melt over rice or dollop over foods before clay-pot baking.*

… Even as the green herb have I given you all things.
—Genesis 9:3

Ghee and Honey

Mix two parts honey to one part ghee. Mix together without heating. This spread is delightful over bagels or toast. Add a pinch of cinnamon for extra warmth and aroma.

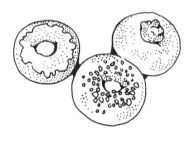

Salt is white and pure—there is something holy in salt.
—Nathaniel Hawthorne

Ghee can replace butter and margarine in almost all your cooking and baking and it enhances the flavor and texture at the same time. Since ghee can be taken to a temperature of 170 degrees without burning or smoking, it exceeds all other oils for the best choice when stir-frying.

For baking or stove-top cooking, the general rule is to use half as much ghee as you would butter. This is a superb trade-off since you are able to use ghee to cut the grams of fat and increase the flavor in most recipes. Using ghee will extend the shelf life of your baked goods by 30%.

In the case of refrigerator cookies or pie dough, it is better to stick with regular butter or a mixture of one-half butter and one-quarter ghee. In this case, the milk fat solids of the butter serve the purpose of binding the dough together and creating a texture that is dense and heavy.

Each week we receive letters from people who have just begun to use ghee and are hooked. The key is to just start using it. As if on cue, people always end their letters saying, "I will never go back to using regular butter again." I hope that will be the case in your kitchen too.

In the back of this book, I have listed the titles of a few of my all-time favorite cook books, each containing recipes using ghee.

Parsley—the jewel of herbs, both in the pot and on the plate.
—Albert Stockli, Twentieth Century Swiss Chef

My friend, Anju Mahaldar, is a master at traditional Indian cuisine. A dinner party at her house includes wonderful vegetarian delicacies like pakoras, samosas and exotic dals and curries.

Anju once told me that her mother had so many delicious dal (Indian pea or lentil soup) recipes that she could make a different one every night for a month.

Anju generously shared these two special recipes with me for this book.

Enjoy!

Non-spicy Split Pea Dal

Ingredients:

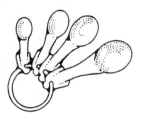

> 1 cup yellow toor dal
> 3 cups water
> 1/2 tsp. tumeric
> 1 tsp. cumin seeds
> salt to taste
> 1-2 tsp. ghee
> 2 tsp. lemon juice

Preparation:

Wash dal thoroughly. Cook with water, tumeric and cumin seeds until soft and smooth. Add salt, lemon juice and ghee. Variation: Instead of lemon juice add 1 tsp. sour cream or 1 tsp. fresh cream for a creamier dal.

This is a very simple recipe, yet Anju's guest will always say, "This is so delicious. What's in this?" She also notes that this particular dal recipe is so light and healthy and easy to digest that it is good to have during convalescence.

Whether therefore ye eat, or drink, or whatsoever ye do, do all to the glory of God.
—Corinthians 10:31

Spicy Mung Dal

Ingredients:

> 1 cup split mung dal
> 3 cups water
> 1 small tomato
> 1/2 cup fresh cilantro leaves
> 1/2 tsp. jalapenos—chopped
> > or
> 1 tsp. sweet bell pepper chopped
> 1/2 tsp. tumeric
> 1/2 inch piece dry mango
> 1/4 tsp. grated ginger
> 1/2 tsp. whole cumin
> 3 dried whole red chiles broken in half,
> > seeds removed
> 1/2 tsp. salt (or to taste)
> 1 tsp. ghee

Preparation:

Wash mung dal thoroughly. Add water, chopped tomatoes, tumeric, dry mango and grated ginger. Bring to a boil. Stir, turn down heat and allow to simmer until dal is soft and blends in with the water. Heat ghee in separate pan, add cumin and red chilies and cook them in the melted ghee until lightly browned. Pour the ghee and spices over the dal. It should sizzle as it is poured. Add cilantro, chopped peppers and salt. Serve immediately while the dal is hot and aromatic.

OUT THE FRYING PAN, INTO THE FIRE
Ghee Candles/Massage Oils/Agnihotra

GHEE CANDLES

In India, Ghee candles are often used at celebrations and sacred ceremonies. Ghee is believed to have purificatory properties and this enhances the effect of the ceremony.

To make ghee candles at home, all you need are some votive candle holders, some wire core wicking, metal wick holder tabs and some melted ghee. I use Walnut Hill Candle Supplies available at most arts and craft stores (see the directory in the back of this book).

Clip your wire covered wick one-half inch longer than the top of the votive candle holder. Wrap it around the wick holder tab and center the tab in the base of the votive. Now gently pour one-quarter cup of ghee into the candle holder, keeping the wick centered.

Light the wick and enjoy the golden glow! One-quarter cup of ghee will burn for nine hours!

I use ghee candles indoors only and observe all the same precautions that I would with any other candles.

For special dinner parties or holiday celebrations you can add a drop of pure lemon or peppermint oil in each votive— it will emit a delicate aroma along with the candle's warm glow.

Incidentally, if you accidentally over cooked your ghee (it has happened to all of us), you will have enough ghee to make several months worth of ghee candles!

MASSAGE OIL

Dana Roehlke, a body worker in Missouri, called us at Purity Farms, Inc. and ordered several buckets of ghee. In our conversation, she mentioned that she used the ghee in her massage therapy practice. She generously shared her ghee-based massage oil with me. Here it is:

> *1/2 cup ghee*
> *1/4 cup jojoba oil*
> *1/2 cup shea butter*

To this mixture, add 13 to 20 drops of pure orange oil with an eye dropper. She said you can also use lavender or rosemary oil in place of orange oil.

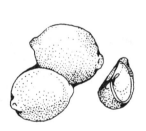

ROSEMARY

AGNIHOTRA

Agnihotra is a fire ceremony that has been practiced in India since ancient times. Ghee is an integral part of this ceremony, as explained below. Richard Powers is an Agnihotra practitioner who explained it to me this way.

Agnihotra is a healing fire from the ancient science of Ayurveda. It is a process of purifying the atmosphere through a specially prepared fire performed at sunrise and sunset daily. Thousands of people have experienced that Agnihotra reduces stress, leads to greater clarity of thought, improves overall health, gives one increased energy and makes the mind more full of love. One essential ingredient used in Agnihotra is ghee. Ghee is a unique substance in that it does not lose weight when it is burned. In Agnihotra, ghee serves as a catalyst which speeds plant metabolism. It creates a chemical reaction with the plant, and the net result is that the plants mature faster and give optimum yield when Agnihotra is performed regularly in the garden. In Agnihotra, ghee attaches itself to the molecular structure of the soil, which facilitates the soil's water retention.

I have listed a company in the back of this book that specializes in Agnihotra supplies and education.

They traded in thy market wheat of Minnith, and Pan-nag,
and honey, and oil, and balm.

—Ezekiel 27:27

HEADING IN THE RIGHT DIRECTION
Sources for Ghee Materials & Related Books

Cookbooks

Ayurveda, A Life in Balance. Maya Tiwari, Healing Arts Press.

A thorough guide to Ayurveda, body type analysis and recipes for balancing the different body types in each season. A real masterpiece.

*Lord   Krishna's   Cuisine: The   Art   of   Inidan
Vegetarian Cooking.*   Yamuna Devi, Bala Books.

Gourmet Magazine called this cookbook "Difinitive" and
it really is the   best Indian vegetarian cookbook I have
found.

The Body Ecology Diet
126 West Paces Ferry Road, Suite # 505
Atlanta, Georgia   30327
Phone: 404-352-8048
A comprehensive approach to diet and health for people
with candida and food allergies.

Walnut Hill Company
P.O. Box # 599
Bristol, Pennsylvania   19007
Candle Making Supplies

ORGANIC DAIRY PRODUCTS

Organic Valley C.R.O.P.P. Cooperative

Main Street

Lafarge, WI 54639

Natural Horizons Dairy

7490 Clubhouse Road

Boulder, CO 80301

AGNIHOTRA INFORMATION AND SUPPLIES

Copper-Works

Route #8 Box 365

Madison, VA 22727

WECK GLASS CANNING JARS AND
WIDE MOUTH STAINLESS STRAINERS

Sur La Table

Kitchen Catalogue

410 Terry Avenue North

Seattle, WA 98109-5229

1 800-243-0852

These canning jars are as practical as they are beautiful and come with glass tops and snap-on clips for air tight sealing.

INFORMATION ON ORGANIC (BGH FREE) DAIRY PRODUCTS

The Pure Food Campaign

(a non-profit organization)

860 Hwy 61

Little Marais, MN 55614

RELATED READING

Ayurveda - The Science of Self Healing by Dr. Vasant Lad, Lotus Press.

The Yoga of Herbs by Dr. Vasant Lad and David Frawley, Lotus Press.

Perfect Health: The Complete Mind/Body Guide, by Dr. Deepak Chopra, M.D., Harmony Books.

AYURVEDIC HEALTH AND REJUVENATION CENTERS

The Ayurvedic Institute and Wellness Center

P.O. Box 23445

Albuquerque, NM 87192-1445

(505) 291-9698

Ayurvedic Holistic Center

82 A Bayville Avenue

Bayville, New York 11709

Swami Sada Shiva Tirtha

Ph: 516-628-8200

Chopra Center For Wellness

7590 Fay Avenue

LaJolla, California 92037

Ph: 619-551-7788

Lotus Ayurvedic Center

4145 Clares Street, Suite D

Capitola, California 95010

Ph: 408-479-1667

California College of Ayurveda

1117 A East Main Street

Grass Valley, California 95945

Ph: 530-274-9100

If You are looking for fine, all natural cosmetics based on Ayurvedic formulas:

Shivani Cosmetics
Devi, Inc.
Lancaster, Mass. 01523
Ph:(Within U.S.) 1-800-237-8221
Ph: (Outside U.S.) 978-365-6532
Website: www.Shivani.com
email: Customer_service@shivani.com

Shivani manufactures a complete line of skin care, hair care, natural color cosmetics and essential oil fragrances suitable for all skin types. All the products are cruelty free & contain no alcohol, mineral oil or artificial fragrances. Every product they manufacture is vegetable based, using 100% natural herbal extracts, essential oils and envoirnmentally safe packaging.

"Like the oil that is in the til seed,

Like the presence of ghee in milk ,

Like the fragrance in a flower,

Like the juice in a fruit,

Like fire in a piece of wood,

So does the Divine permeate the cosmos"

-Sri Sathya Sai Baba, Sanathana Sarathi

I SAW IT THERE FIRST
Bibliography

Atherton, Henry, and Newlander, John A. *Chemistry and Testing of Dairy Products.* 4th ed. AVI 1977.

AU. "Antibiotics: A Medical Disaster in the Making," *Grassroots* (Newspaper of Pure Food Campaign), October 1995.

AU. *CHARAK SAMHITA.* Varanasi, India: Chowkhamba Sanskrit Series, 1977.

Ballentine, Rudolph, Dr. *TRANSITION TO VEGETARIANISM: An Evolutionary Step.* Henesdale, Pennslyvania: Himalayan International Institute of Yoga Science, 1987.

Chopra, Deepak, MD. *PERFECT HEALTH: The Complete Mind/Body Guide.* New York: Harmony Books, 1991.

Finnegan, John, N.D. *The Facts About Fats: A Consumers Guide to Good Oils.* Berkeley: Celestial Arts Publishing, 1993.

Gates, Donna & Schatz, Linda. *The Body Ecology Diet,* Atlanta: B.E.D. Publications, 1993.

Lad, Vasant Dr. *AYURVEDA: The Science of Healing.* Wisconsin: Lotus Press, 1984.

Lampert, Lincoln M. *Modern Dairy Products.* 3rd ed. Chemical Publishing 1975.

National Dairy Council. *How Americans Use their Dairy Foods.* 1967.
Powell, Elizabeth A. *Pennsylvania Butter: Tools and Processes.* Bucks Co. Hist. 1974.

Schneider, Keith. "Grocers Challenge Use of New Drug For Milk Output" *The New York Times* 4 February 1994:

Tiwari, Maya. *AYURVEDA: A Life of Balance. Rochester,* Vermont: Healing Arts Press, 1995.

We recommend you cook with ghee made at home or by Purity Farms. Ghee is less mucous-forming than butter and contains no lactose (milk sugar), so it is ideal for an anti-candida diet.
—*Donna Gates, Author*

Spice a dish with love and it pleases every palate.

—*Plautus*

NOTES

NOTES

To order additional copies of *Ghee: A Guide to the Royal Oil*, complete the information below:

Ship to: (please print)

Name _____

Address _____

City, State, Zip _____

Day phone _____

_____copies of Ghee...@$ 9.95* each $_____

Postage & handling @ $2.50 each $_____

Colorado residents add 3% tax $_____

Total amount enclosed $_____

Make checks payable to Ghee...

Send to: Kathryn S. Feldenkreis

14635 Westcreek Road • Sedalia, CO 80135

*price may change

— —

To order additional copies of Ghee: A Guide to the Royal Oil, complete the information below:

Ship to: (please print)

Name _____

Address _____

City, State, Zip _____

Day phone _____

_____copies of Ghee...@$ 9.95* each $_____

Postage & handling @ $2.50 each $_____

Colorado residents add 3% tax $_____

Total amount enclosed $_____

Make checks payable to Ghee...

Send to: Kathryn S. Feldenkreis

14635 Westcreek Road • Sedalia, CO 80135

*price may change